Read All About

THE

UNIVERSE

by Lucy Beevor

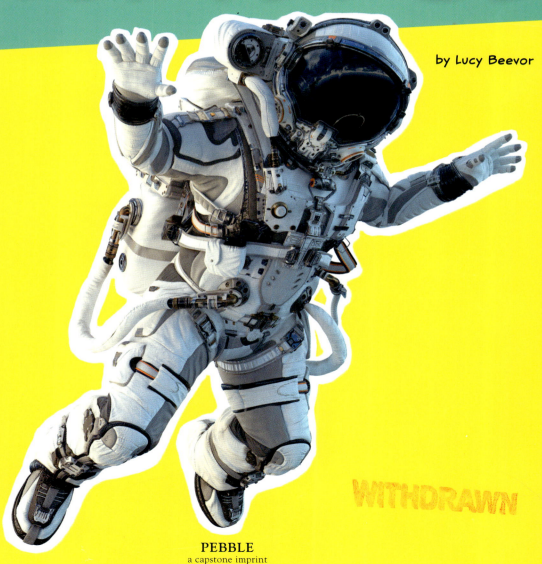

PEBBLE
a capstone imprint

Pebble Sprout is published by Pebble, an imprint of Capstone.
1710 Roe Crest Drive
North Mankato, Minnesota 56003
www.capstonepub.com

Library of Congress Cataloging-in-Publication Data is available on the Library of Congress website.
ISBN: 978-1-9771-3225-3 (library binding)
ISBN: 978-1-9771-5481-1 (ebook pdf)

Summary: Just how big is the universe? Find out all about the universe and its planets, stars, and more in this fact-filled book. It's perfectly designed to introduce young children to the power of nonfiction. Stunning art and photos give readers a fascinating look at the wonders of the universe, from the earth, moon, sun, and beyond!

Image Credits
Alamy: NASA Photo, top right 29; NASA, 23, bottom 31, Carissa Callini, middle 29; Shutterstock: 19 STUDIO, 15, 3Dsculptor, 28, Allexxandar, bottom 5, top 9, 13, anatoliy_gleb, 10, Andrey Armyagov, top left 6, 7, Denis Belitsky, 8, Dotted Yeti, 30, e71lena, top 31, faboi, bottom 25, Fotos593, middle 17, HelenField, bottom left 22, Juergen Faelchle, bottom 27, Katrina Leigh, bottom 11, MarcelClemens, 19, bottom 20, NASA images, bottom left 6, 9, 16, top 18, NikoNomad, (bottom) Cover, ninanaina, (top) Cover, osaka.maki, bottom left 12, Paopano, 26, Paul Fleet, 24, Quality Stock Arts, 4, Robert Eder Astronomy, middle left 9, Sergey Nivens, (top right) Cover, 1, sNike, top 11, solarseven, top 25, taffpixture, top right 22, Triff, top right 12, top right 27, Vadim Sadovski, top 5, top 17, bottom 18, top 20, top 21, bottom 21, bottom 23, Withan Tor, 14; Wikimedia: NASA, bottom left 29

Editorial Credits
Editor: Peter Mavrikis; Designer: Kayla Rossow; Media Researcher: Morgan Walters; Production Specialist: Laura Manthe

Table of Contents

Words in **bold** are in the glossary.

Chapter 1

What Is the Universe?

The universe is everything we can see, and all that we cannot see. Are you ready to launch into space to find out more? Three . . . two . . . one . . . Liftoff!

The universe started 14 billion years ago with a huge explosion called the Big Bang.

Tiny **particles** from the Big Bang formed stars, planets, and galaxies.

We do not know how big the universe is. Some scientists think it goes on forever.

There is no air in space. Astronauts wear spacesuits to help them breathe.

No one can hear you talk in space. It is totally silent.

There is very little **gravity** in space. Astronauts can float there.

Chapter 2

Galaxies

Planet Earth is part of our solar system.
Our solar system is a tiny part of a galaxy
called the Milky Way. But what is a galaxy?

A galaxy is a huge cloud of dust, stars, and planets.

The Milky Way

There could be 100 to 200 billion galaxies in the universe.

The Andromeda galaxy is bright enough to be seen without a **telescope** on clear, dark nights at certain times of the year.

Galaxies come in four basic shapes. Some are odd shapes. The Tadpole galaxy looks like a giant tadpole!

Chapter 3

Stars

Look up at the sky on a clear night. What can you see? If you're lucky, you will see thousands of shiny stars.

Stars twinkle like fairy lights. But they are really giant balls of hot, glowing gas.

There are about 100 to 400 billion stars in the Milky Way. That's a lot of stars!

The nearest star to Earth is the sun. The sun is a type of star called a yellow dwarf.

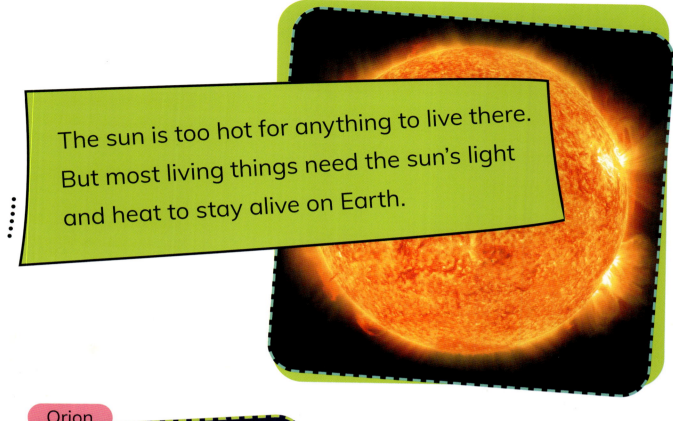

The sun is too hot for anything to live there. But most living things need the sun's light and heat to stay alive on Earth.

Orion

Constellations are groups of stars. When you "connect the stars," they make a picture in the sky.

The Big Dipper is part of the constellation Ursa Major. What do you think it looks like?

Chapter 4

Our Solar System

You may think that Earth is a very big place. But it is only a tiny part of our solar system.

The sun is in the middle of our solar system. All other objects in the solar system move around the sun.

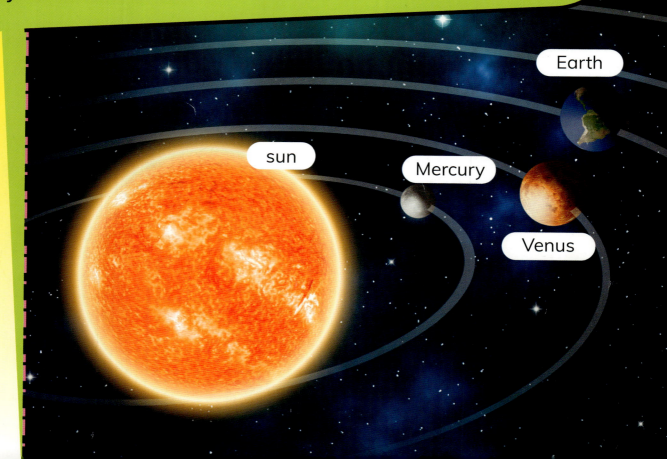

sun

Mercury

Venus

Earth

It takes 365 days for Earth to **orbit** the sun. This is called a year.

Earth

sun

Uranus

Jupiter

Neptune

Saturn

Mars

Neptune's orbit is the longest. It lasts for 165 years!

Chapter 5

Planets

There are eight planets in our solar system, including Earth. The planets are all very different worlds.

Mercury is the smallest planet in our solar system. It is also the closest planet to the sun.

Venus is the hottest planet in the solar system. It is so hot it could melt metal!

Earth is the only planet in our solar system known to have living things. It looks like a giant blue and green marble from space.

Mars is called the "Red Planet" because the rocky ground is covered in red **rust**.

Jupiter is by far the biggest planet. About 1,321 Earths could fit inside Jupiter!

Saturn has thousands of rings around its middle. The rings are made of chunks of sparkly ice and dust.

Uranus is an "ice giant" planet. It is the coldest planet in our solar system.

Neptune is the farthest planet from the sun. It took almost nine years for one **spacecraft** to reach Neptune from Earth.

Pluto was the ninth planet. But scientists decided it was too small to be a planet. Now Pluto is a "dwarf planet."

Exoplanets are planets far outside our solar system. There could even be life on these faraway planets!

Chapter 6

Moons

Every night, the sun sets, and the sky goes dark. Then a big round ball rises bright in the sky. It is the moon, Earth's closest neighbor.

The moon doesn't make its own light. It **reflects** light from the sun.

mountain

crater

sea

The moon is a dusty gray rock with deep **craters**, tall mountains, and flat parts called "seas."

Twelve astronauts have walked on the moon. The first was Neil Armstrong in 1969.

Neil Armstrong

Earth's moon isn't the only moon. We know of more than 200 moons in our solar system.

Saturn has 82 moons! Scientists are finding more all the time.

Chapter 7

Asteroids, Comets, and Meteors

Thousands of space rocks orbit the sun. They are called asteroids, comets, and meteors.

Asteroids are great big chunks of rock or metal. They are left over from when the planets formed.

Comets are giant snowballs made of ice, gas, and dust.

Shooting stars aren't really stars. They are rocks called meteors. They shoot across the sky, leaving a glowing trail behind them.

Chapter 8

Exploring the Universe

Astronomers, astronauts, robots, and telescopes help us to learn about space. So far only 566 people have been to space.

Many astronauts flew to space in the space shuttle. A new spacecraft called *Dragon* took astronauts to space in 2020.

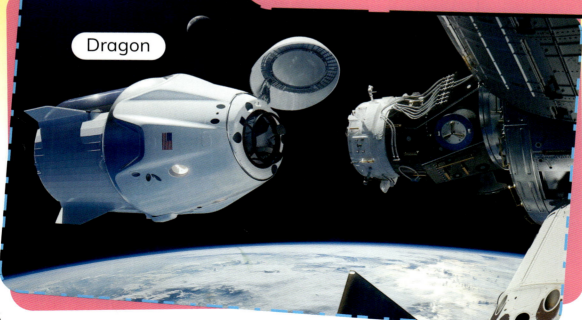

Dragon

Mars rovers are robots on wheels. Scientists send them to Mars to search for water and other signs of life.

The Hubble Space Telescope takes photographs of stars and planets from space.

Chapter 9

Living and Working in Space

Living in space is very different from living on Earth. Astronauts have to learn new ways to work, eat, and sleep.

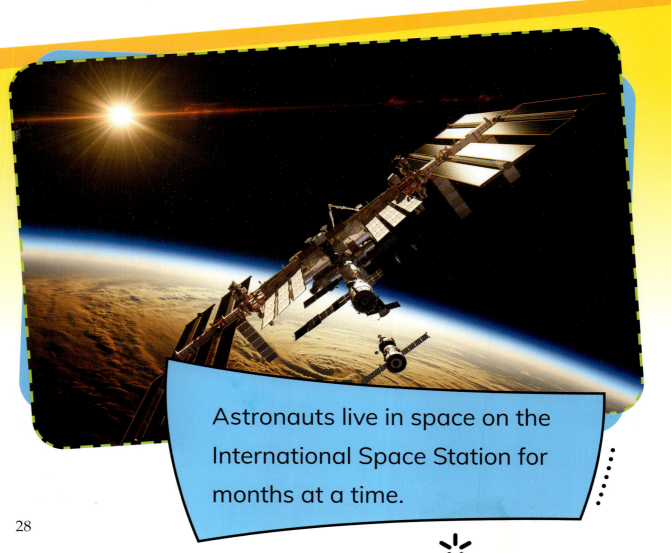

Astronauts live in space on the International Space Station for months at a time.

Scientists run many experiments onboard the Space Station.

Astronauts must strap into their beds. This keeps them from floating around while they sleep.

Most space food comes in packets and is **freeze-dried**. Hot water must be added before the food can be eaten.

To Infinity . . . *and Beyond!*

The universe is a very big place. There is still so much to discover. One day, ordinary people might live and work in space too!

3D illustraton

Scientists plan to build houses on the moon. Astronauts could live there for months at a time.

3D illustraton

People could one day live on Mars. Air will be pumped into buildings to help them breathe.

Orion is a new spacecraft. It will take people deeper into space than ever before.

Glossary

crater—a big hole in the ground made by a meteor

freeze-dried—food that is frozen and dried so it stays fresh

gravity—the force that pulls things down and keeps people from floating away into space

orbit—the path an object takes as it moves around the sun

particle—a very tiny piece that joins together with other tiny pieces to make the objects we see around us

reflect—when light bounces off the surface of an object

rust—the reddish-brown layer that forms on metals and some rocks

spacecraft—a vehicle used for space travel

telescope—tool used to see objects that are very far away, like stars and planets

Index